Wilma Rudolph was born undersized and underweight. She remained unhealthy while growing up in her family's home in Tennessee. Her overworked mother nursed her through many diseases. The worst came when Wilma's left leg grew weak and disabled. Wilma learned that she had polio (pō´ lē ō), a disease that affects the muscles. The doctor said Wilma would be unable to walk again.

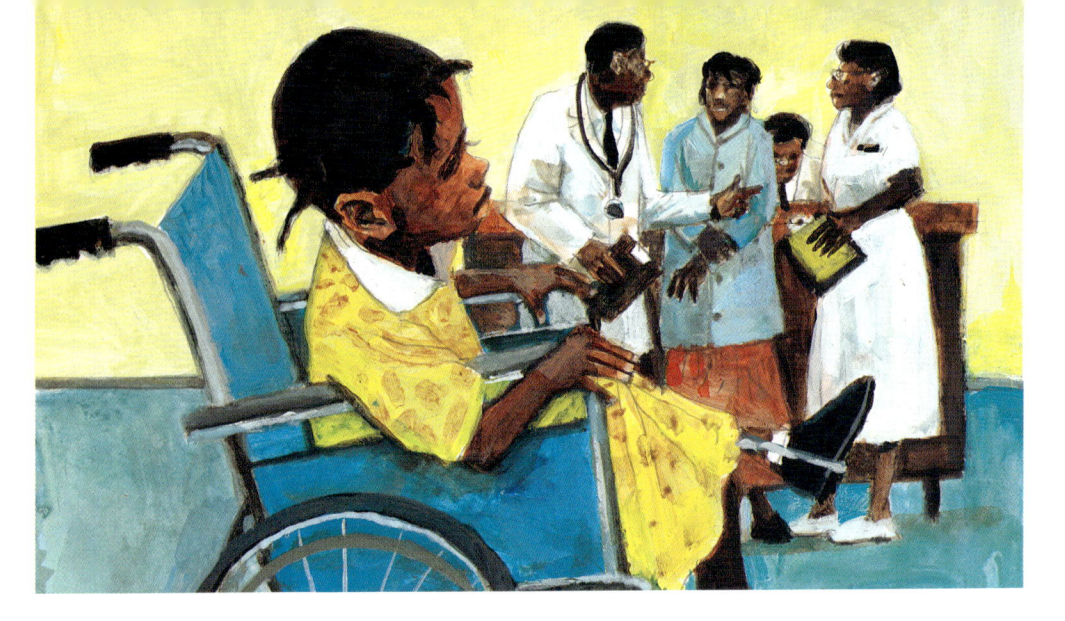

The Rudolphs were unwilling to accept the unhappy news. They prepared to do anything they could to help their sick daughter. Mrs. Rudolph took Wilma fifty miles to a hospital for African-Americans. (At that time, African-Americans in the south could not use the same hospitals as whites.) Mrs. Rudolph was overjoyed when doctors said they might be able to help Wilma regain her walking.

Wilma's road to recovery was difficult but not impossible. She returned to the hospital twice a week for two years. During her biweekly visits, she was shown how to walk with a metal leg brace. She was also given home exercises to rebuild the strength in her leg. Wilma repeated the same, monotonous exercises day after day, trying to overcome the weakness in her leg muscles.

The Rudolph family included Wilma's 21 brothers and sisters. All of them joined to help Wilma become as independent as possible. They cheered her on as she did her tough, indoor exercises. Over the years, Wilma's pain and discomfort became less and less. Finally, at age 12, she was able to walk again without braces, crutches, or canes.

The family was overjoyed that Wilma had overcome her weakness. But they were in disbelief when she announced her next goal to undertake: Wilma wanted to be an athlete!

True to her word, Wilma became a high school basketball star. She set state records for scoring. Clearly, Wilma and her goals were not to be mistrusted!

In college, Wilma joined the women's track team. Her speed and spirit impressed the public. They found Wilma irresistible. Her fame quickly spread. Wilma was invited to run in track events around the world. Soon the number of invitations tripled (trĭp´ əld) and then quadrupled. Wilma left school for a year before reentering to finish her studies.

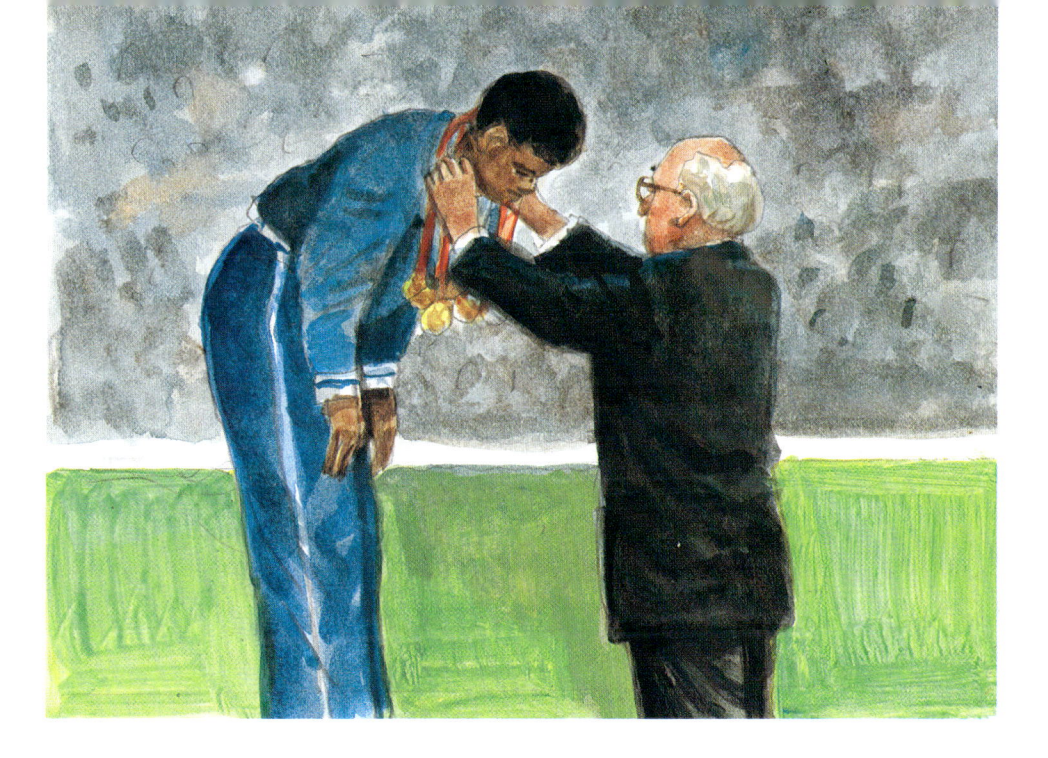

In 1960, Wilma became a hero unlike any other. She competed in the Olympics in Rome, Italy. Wilma won a trio (trē´ ō) of track events: the 100-meter dash, the 200-meter dash, and the 400-meter relay. Wilma Rudolph became the first woman in America ever to win three Olympic gold medals.

After college, Wilma returned home to be a teacher and track coach. She was invited to speak at many schools. She restated her message often: Don't prejudge people because of their physical condition. Someone you misjudge as weak might just win a pentathlon (a track-and-field contest with five events). Wilma Rudolph showed that, with a strong will and spirit, you can overcome difficult problems.